MR. GLADSTONE AND GENESIS
ESSAY #5 FROM "SCIENCE AND HEBREW TRADITION"

By Thomas Henry Huxley

Contents

NOTE ON THE PROPER SENSE OF THE "MOSAIC" NARRATIVE OF THE CREATION.
FOOTNOTES

In controversy, as in courtship, the good old rule to be off with the old before one is on with the new, greatly commends itself to my sense of expediency. And, therefore, it appears to me desirable that I should preface such observations as I may have to offer upon the cloud of arguments (the relevancy of which to the issue which I had ventured to raise is not always obvious) put forth by Mr. Gladstone in the January number of this review, 1 by an endeavour to make clear to such of our readers as have not had the advantage of a forensic education the present net result of the discussion.

I am quite aware that, in undertaking this task, I run all the risks to which the man who presumes to deal judicially with his own cause is liable. But it is exactly because I do not shun that risk, but, rather, earnestly desire to be judged by him who cometh after me, provided that he has the knowledge and impartiality appropriate to a judge, that I adopt my present course.

In the article on "The Dawn of Creation and Worship," it will be remembered that Mr. Gladstone unreservedly commits himself to three propositions. The first is that, according to the writer of the Pentateuch, the "water-population," the "air-population," and the "land-population" of the globe were created successively, in the order named. In the second place, Mr. Gladstone authoritatively asserts that this (as part of his "fourfold order") has been "so affirmed in our time by natural science, that it may be taken as a demonstrated conclusion and established fact." In the third place, Mr. Gladstone argues that the fact of this coincidence of the pentateuchal story with the results of modern investigation makes it "impossible to avoid the conclusion, first, that either this writer was gifted with faculties passing all human experience, or else his knowledge was divine." And having settled to his own satisfaction that the first "branch of the alternative is truly nominal and unreal," Mr. Gladstone continues, "So stands the plea for a revelation of truth from God, a plea only to be met by questioning its possibility" (p. 697).

I am a simple-minded person, wholly devoid of subtlety of intellect, so that I willingly admit that there may be depths of alternative meaning in these propositions out of all soundings attainable by my poor plummet. Still there are a good many people who suffer under a like intellectual limitation; and, for once in my life, I feel that I have the chance of attaining that position of a representative of average opinion which appears to be the modern ideal of a leader of men, when I make free confession that, after turning the matter over in my mind, with all the aid derived from a careful consideration of Mr. Gladstone's reply, I cannot get away from my original conviction that, if Mr. Gladstone's second proposition can be shown to be not merely inaccurate, but directly contradictory of facts known to every one who is acquainted with the elements of natural science, the third proposition collapses of itself.

And it was this conviction which led me to enter upon the present discussion. I fancied that if my respected clients, the people of average opinion and capacity, could once be got distinctly to conceive that Mr. Gladstone's views as to the proper method of dealing with grave and difficult scientific and religious problems had permitted him to base a solemn "plea for a revelation of truth from God" upon an error as to a matter of fact, from which the intelligent perusal of a manual of palaeontology would have saved him, I need not trouble myself to occupy their time and attention with further comments upon his contribution to apologetic literature. It is for others to judge whether I have efficiently carried out my project or not. It certainly does not count for much that I should be unable to find any flaw in my own case, but I think it counts for a good deal that Mr. Gladstone appears to have been equally unable to do so. He does, indeed, make a great parade of authorities, and I have the greatest respect for those authorities whom Mr. Gladstone mentions. If he will get them to sign a joint memorial to the effect that our present palaeontological evidence proves that birds appeared before the "land-population" of terrestrial reptiles, I shall think it my duty to reconsider my position—but not till then.

It will be observed that I have cautiously used the word "appears" in referring to what seems to me to be absence of any real answer to my criticisms in Mr. Gladstone's reply. For I must honestly confess that, notwithstanding long and painful strivings after clear insight, I am still uncertain whether Mr. Gladstone's "Defence" means that the great "plea for a revelation from God" is to be left to perish in the dialectic desert; or whether it is to be withdrawn under the protection of such skirmishers as are available for covering retreat.

In particular, the remarkable disquisition which covers pages 11 to 14 of Mr. Gladstone's last contribution has greatly exercised my mind. Socrates is reported to have said of the works of Heraclitus that he who attempted to comprehend them should be a "Delian swimmer," but that, for his part, what he could understand was so good that he was disposed to believe in the excellence of that which he found unintelligible. In endeavouring to make myself master of Mr. Gladstone's meaning in these pages, I have often been overcome by a feeling analogous to that of Socrates, but not quite the same. That which I do understand has appeared to me so very much the reverse of good, that I have sometimes permitted myself to doubt the value of that which I do not understand.

In this part of Mr. Gladstone's reply, in fact, I find nothing of which the bearing upon my arguments is clear to me, except that which relates to the question whether reptiles, so far as they are represented by tortoises and the great majority of lizards and snakes, which are land animals, are creeping things in the sense of the pentateuchal writer or not.

I have every respect for the singer of the Song of the Three Children (whoever he may have been); I desire to cast no shadow of doubt upon, but, on the contrary, marvel at, the exactness of Mr. Gladstone's information as to the considerations which "affected the method of the Mosaic writer"; nor do I venture to doubt that the inconvenient intrusion of these contemptible reptiles—"a family fallen from greatness" (p. 14), a miserable decayed aristocracy reduced to mere "skulkers about the earth" (*ibid.*)—in consequence, apparently, of difficulties about the occupation of land arising out of the earth-hunger of their former serfs, the mammals—into an apologetic argument, which otherwise would run quite smoothly, is in every way to be deprecated. Still, the wretched creatures stand there, importunately demanding notice; and, however different may be the practice in that contentious atmosphere with which Mr. Gladstone expresses and laments his familiarity, in the atmosphere of science it really is of no avail whatever to shut one's eyes to facts, or to try to bury them out of sight under a tumulus of rhetoric. That is my experience of the "Elysian regions of Science," wherein it is a pleasure to me to think that a man of Mr. Gladstone's intimate knowledge of English life, during the last quarter of a century, believes my philosophic existence to have been rounded off in unbroken equanimity.

However reprehensible, and indeed contemptible, terrestrial reptiles may be, the only question which appears to me to be relevant to my argument is whether these creatures are or are not comprised under the denomination of "everything that creepeth upon the ground."

Mr. Gladstone speaks of the author of the first chapter of Genesis as "the Mosaic writer"; I suppose, therefore, that he will admit that it is equally proper to speak of the author of Leviticus as the "Mosaic writer." Whether such a phrase would be used by any one who had an adequate conception of the assured results of modern Biblical criticism is another matter; but, at any rate, it cannot be denied that Leviticus has as much claim to Mosaic authorship as Genesis. Therefore, if one wants to know the sense of a phrase used in Genesis, it will be well to see what Leviticus has to say on the matter. Hence, I commend the following extract from the eleventh chapter of Leviticus to Mr. Gladstone's serious attention:—

And these are they which are unclean unto you among the creeping things that creep upon the earth: the weasel, and the mouse, and the great lizard after its kind, and the gecko, and the land crocodile, and the sand-lizard, and the chameleon. These are they which are unclean to you among all that creep (v. 29-31).

The merest Sunday-school exegesis therefore suffices to prove that when the "Mosaic writer" in Genesis i. 24 speaks of "creeping things," he means to include lizards among them.

This being so, it is agreed, on all hands, that terrestrial lizards, and other reptiles allied to lizards, occur in the Permian strata. It is further agreed that the Triassic strata were deposited after these. Moreover, it is well known that, even if certain footprints are to be taken as unquestionable evidence of the existence of birds, they are not known to occur in rocks earlier than the Trias, while indubitable remains of birds are to be met with only much later. Hence it follows that natural science does not "affirm" the statement that birds were made on the fifth day, and "everything that creepeth on the ground" on the sixth, on which Mr. Gladstone rests his order; for, as is shown by Leviticus, the "Mosaic writer" includes lizards among his "creeping things."

Perhaps I have given myself superfluous trouble in the preceding argument, for I find that Mr. Gladstone is willing to assume (he does not say to admit) that the statement in the text of Genesis as to reptiles cannot "in all points be sustained" (p. 16). But my position is that it cannot be sustained in any point, so that, after all, it has perhaps been as well to go over the evidence again. And then Mr. Gladstone proceeds as if nothing had happened to tell us that—

There remain great unshaken facts to be weighed. First, the fact that such a record should have been made at all.

As most peoples have their cosmogonies, this "fact" does not strike me as having much value.

Secondly, the fact that, instead of dwelling in generalities, it has placed itself under the severe conditions of a chronological order reaching from the first nisus of chaotic matter to the consummated production of a fair and goodly, a furnished and a peopled world.

This "fact" can be regarded as of value only by ignoring the fact demonstrated in my previous paper, that natural science does not confirm the order asserted so far as living things are concerned; and by upsetting a fact to be brought to light presently, to wit, that, in regard to the rest of the pentateuchal cosmogony, prudent science has very little to say one way or the other.

Thirdly, the fact that its cosmogony seems, in the light of the nineteenth century, to draw more and more of countenance from the best natural philosophy.

I have already questioned the accuracy of this statement, and I do not observe that mere repetition adds to its value.

And, fourthly, that it has described the successive origins of the five great categories of present life with which human experience was and is conversant, in that order which geological authority confirms.

By comparison with a sentence on page 14, in which a fivefold order is substituted for the "fourfold order," on which the "plea for revelation" was originally founded, it appears that these five categories are "plants, fishes, birds, mammals, and man," which, Mr. Gladstone affirms, "are given to us in Genesis in the order of succession in which they are also given by the latest geological authorities."

I must venture to demur to this statement. I showed, in my previous paper, that there is no reason to doubt that the term "great sea monster" (used in Gen. i. 21) includes the most conspicuous of great sea animals—namely, whales, dolphins, porpoises, manatees, and dugongs;2 and, as these are indubitable mammals, it is impossible to affirm that mammals come after birds, which are said to have been created on the same day. Moreover, I pointed out that as these Cetacea and Sirenia are certainly modified land animals, their existence implies the antecedent existence of land mammals.

Furthermore, I have to remark that the term "fishes," as used, technically, in zoology, by no means covers all the moving creatures that have life, which are bidden to "fill the waters in the seas" (Gen. i. 20-22.) Marine mollusks and crustacea, echinoderms, corals, and foraminifera are not technically fishes. But they are abundant in the palaeozoic rocks, ages upon ages older than those in which the first evidences of true fishes appear. And if, in a geological book, Mr. Gladstone finds the quite true statement that plants appeared before fishes, it is only by a complete misunderstanding that he can be led to imagine it serves his purpose. As a matter of fact, at the present moment, it is a question whether, on the bare evidence afforded by fossils, the marine creeping thing or the marine plant has the seniority. No cautious palaeontologist would express a decided opinion on the matter. But, if we are to read the pentateuchal statement as a scientific document (and, in spite of all protests to the contrary, those who bring it into comparison with science do seek to make a scientific document of it), then, as it is quite clear that only terrestrial plants of high organisation are spoken of in verses 11 and 12, no palaeontologist would hesitate to say that, at present, the records of sea animal life are vastly older than those of any land plant describable as "grass, herb yielding seed or fruit tree."

Thus, although, in Mr. Gladstone's "Defence," the "old order passeth into new," his case is not improved. The fivefold order is no more "affirmed in our time by natural science" to be "a demonstrated conclusion and established fact" than the fourfold order was. Natural science appears to me to decline to have anything to do with either; they are as wrong in detail as they are mistaken in principle.

There is another change of position, the value of which is not so apparent to me, as it may well seem to be to those who are unfamiliar with the subject under discussion. Mr. Gladstone discards his three groups of "water-population," "air-population," and "land-population," and substitutes for them (1) fishes, (2) birds, (3) mammals, (4) man. Moreover, it is assumed, in a note, that "the higher or ordinary mammals" alone were known to the "Mosaic writer" (p. 6). No doubt it looks, at first, as if something were gained by this alteration; for, as I have just pointed out, the word "fishes" can be used in two senses, one of which has a deceptive appearance of adjustability to the "Mosaic" account. Then the inconvenient reptiles are banished out of sight; and, finally, the question of the exact meaning of "higher" and "ordinary" in the case of mammals opens up the prospect of a hopeful logomachy. But what is the good of it all in the face of Leviticus on the one hand and of palaeontology on the other?

As, in my apprehension, there is not a shadow of justification for the suggestion that when the pentateuchal writer says "fowl" he excludes bats (which, as we shall see directly, are expressly included under "fowl" in Leviticus), and as I have already shown that he demonstrably includes reptiles, as well as mammals, among the creeping things of the land, I may be permitted to spare my readers further discussion of the "fivefold order." On the whole, it is seen to be rather more inconsistent with Genesis than its fourfold predecessor.

But I have yet a fresh order to face. Mr. Gladstone (p. 11) understands "the main statements of Genesis" in successive order of time, but without any measurement of its divisions, to be as follows:—

1. A period of land, anterior to all life (v. 9, 10). 2. A period of vegetable life, anterior to animal life (v. 11, 12). 3. A period of animal life, in the order of fishes (v. 20). 4. Another stage of animal life, in the order of birds. 5. Another in the order of beasts (v. 24, 25). 6. Last of all, man (v. 26, 27).

Mr. Gladstone then tries to find the proof of the occurrence of a similar succession in sundry excellent works on geology.

I am really grieved to be obliged to say that this third (or is it fourth?) modification of the foundation of the "plea for revelation" originally set forth, satisfies me as little as any of its predecessors.

For, in the first place, I cannot accept the assertion that this order is to be found in Genesis. With respect to No. 5, for example, I hold, as I have already said, that "great sea monsters" includes the Cetacea, in which case mammals (which is what, I suppose, Mr. Gladstone means by "beasts") come in under head No. 3, and not under No. 5. Again, "fowl" are said in Genesis to be created on the same day as fishes; therefore I cannot accept an order which makes birds succeed fishes. Once more, as it is quite certain that the term "fowl" includes the bats,—for in Leviticus xi. 13-19 we read, "And these shall ye have in abomination among the fowls... the heron after its kind, and the hoopoe, and the bat,"—it is obvious that bats are also said to have been created at stage No. 3. And as bats are mammals, and their existence obviously presupposes that of terrestrial "beasts," it is quite clear that the latter could not have first appeared as No. 5. I need not repeat my reasons for doubting whether man came "last of all."

As the latter half of Mr. Gladstone's sixfold order thus shows itself to be wholly unauthorised by, and inconsistent with, the plain language of the Pentateuch, I might decline to discuss the admissibility of its former half.

But I will add one or two remarks on this point also. Does Mr. Gladstone mean to say that in any of the works he has cited, or indeed anywhere else, he can find scientific warranty for the assertion that there was a period of land—by which I suppose he means dry land (for submerged land must needs be as old as the separate existence of the sea)—"anterior to all life?"

It may be so, or it may not be so; but where is the evidence which would justify any one in making a positive assertion on the subject? What competent palaeontologist will affirm, at this present moment, that he knows anything about the period at which life originated, or will assert more than the extreme probability that such origin was a long way antecedent to any traces of life at present known? What physical geologist will affirm that he knows when dry land began to exist, or will say more than that it was probably very much earlier than any extant direct evidence of terrestrial conditions indicates?

I think I know pretty well the answers which the authorities quoted by Mr. Gladstone would give to these questions; but I leave it to them to give them if they think fit.

If I ventured to speculate on the matter at all, I should say it is by no means certain that sea is older than dry land, inasmuch as a solid terrestrial surface may very well have existed before the earth was cool enough to allow of the existence of fluid water. And, in this case, dry land may have existed before the sea. As to the first appearance of life, the whole argument of analogy, whatever it may be worth in such a case, is in favour of the absence of living beings until long after the hot water seas had constituted themselves; and of the subsequent appearance of aquatic before terrestrial forms of life. But whether these "protoplasts" would, if we could examine them, be reckoned among the lowest microscopic algae, or fungi; or among those doubtful organisms which lie in the debatable land between animals and plants, is, in my judgment, a question on which a prudent biologist will reserve his opinion.

I think that I have now disposed of those parts of Mr. Gladstone's defence in which I seem to discover a design to rescue his solemn "plea for revelation." But a great deal of the "Proem to Genesis" remains which I would gladly pass over in silence, were such a course consistent with the respect due to so distinguished a champion of the "reconcilers."

I hope that my clients—the people of average opinions—have by this time some confidence in me; for when I tell them that, after all, Mr. Gladstone is of opinion that the "Mosaic record" was meant to give moral, and not scientific, instruction to those for whom it was written, they may be disposed to think that I must be misleading them. But let them listen further to what Mr. Gladstone says in a compendious but not exactly correct statement respecting my opinions:—

He holds the writer responsible for scientific precision: I look for nothing of the kind, but assign to him a statement general, which admits exceptions; popular, which aims mainly at producing moral impression; summary, which cannot but be open to more or less of criticism of detail. He thinks it is a lecture. I think it is a sermon. (p. 5).

I note, incidentally, that Mr. Gladstone appears to consider that the *differentia* between a lecture and a sermon is, that the former, so far as it deals with matters of fact, may be taken seriously, as meaning exactly what it says, while a sermon may not.

I have quite enough on my hands without taking up the cudgels for the clergy, who will probably find Mr. Gladstone's definition unflattering.

But I am diverging from my proper business, which is to say that I have given no ground for the ascription of these opinions; and that, as a matter of fact, I do not hold them and never have held them. It is Mr. Gladstone, and not I, who will have it that the pentateuchal cosmogony is to be taken as science.

My belief, on the contrary, is, and long has been, that the pentateuchal story of the creation is simply a myth. I suppose it to be an hypothesis respecting the origin of the universe which some ancient thinker found himself able to reconcile with his knowledge, or what he thought was knowledge, of the nature of things, and therefore assumed to be true. As such, I hold it to be not merely an interesting, but a venerable, monument of a stage in the mental progress of mankind; and I find it difficult to suppose that any one who is acquainted with the cosmogonies of other nations—and especially with those of the Egyptians and the Babylonians, with whom the Israelites were in such frequent and intimate communication—should consider it to possess either more, or less, scientific importance than may be allotted to these.

Mr. Gladstone's definition of a sermon permits me to suspect that he may not see much difference between that form of discourse and what I call a myth; and I hope it may be something more than the slowness of apprehension, to which I have confessed, which leads me to imagine that a statement which is "general" but "admits exceptions," which is "popular" and "aims mainly at producing moral impression," "summary" and therefore open to "criticism of detail," amounts to a myth, or perhaps less than a myth. Put algebraically, it comes to this, $x=a+b+c$, always remembering that there is nothing to show the exact value of either a, or b, or c. It is true that a is commonly supposed to equal 10, but there are exceptions, and these may reduce it to 8, or 3, or 0; b also popularly means 10, but being chiefly used by the algebraist as a "moral" value, you cannot do much with it in the addition or subtraction of mathematical values; c also is quite "summary," and if you go into the details of which it is made up, many of them may be wrong, and their sum total equal to 0, or even to a minus quantity.

Mr. Gladstone appears to wish that I should (1) enter upon a sort of essay competition with the author of the pentateuchal cosmogony; (2) that I should make a further statement about some elementary facts in the history of Indian and Greek philosophy; and (3) that I should show cause for my hesitation in accepting the assertion that Genesis is supported, at any rate to the extent of the first two verses, by the nebular hypothesis.

A certain sense of humour prevents me from accepting the first invitation. I would as soon attempt to put Hamlet's soliloquy into a more scientific shape. But if I supposed the "Mosaic writer" to be inspired, as Mr. Gladstone does, it would not be consistent with my notions of respect for the Supreme Being to imagine Him unable to frame a form of words which should accurately, or, at least, not inaccurately, express His own meaning. It is sometimes said that, had the statements contained in the first chapter of Genesis been scientifically true, they would have been unintelligible to ignorant people; but how is the matter mended if, being scientifically untrue, they must needs be rejected by instructed people?

With respect to the second suggestion, it would be presumptuous in me to pretend to instruct Mr. Gladstone in matters which lie as much within the province of Literature and History as in that of Science; but if any one desirous of further knowledge will be so good as to turn to that most excellent and by no means recondite source of information, the "Encyclopaedia Britannica," he will find, under the letter E, the word "Evolution," and a long article on that subject. Now, I do not recommend him to read the first half of the article; but the second half, by my friend Mr. Sully, is really very good. He will there find it said that in some of the philosophies of ancient India, the idea of evolution is clearly expressed: "Brahma is conceived as the eternal self-existent being, which, on its material side, unfolds itself to the world by gradually condensing itself to material objects through the gradations of ether, fire, water, earth, and other elements." And again: "In the later system of emanation of Sankhya there is a more marked approach to a materialistic doctrine of evolution." What little knowledge I have of the matter—chiefly derived from that very instructive book, "Die Religion des Buddha," by C. F. Koeppen, supplemented by Hardy's interesting works—leads me to think that Mr. Sully might have spoken much more strongly as to the evolutionary character of Indian philosophy, and especially of that of the Buddhists. But the question is too large to be dealt with incidentally.

And, with respect to early Greek philosophy, 3 the seeker after additional enlightenment need go no further than the same excellent storehouse of information:—

The early Ionian physicists, including Thales, Anaximander, and Anaximenes, seek to explain the world as generated out of a primordial matter which is at the same time the universal support of things. This substance is endowed with a generative or transmutative force by virtue of which it passes into a succession of forms. They thus resemble modern evolutionists since they regard the world, with its infinite variety of forms, as issuing from a simple mode of matter.

Further on, Mr. Sully remarks that "Heraclitus deserves a prominent place in the history of the idea of evolution," and he states, with perfect justice, that Heraclitus has foreshadowed some of the special peculiarities of Mr. Darwin's views. It is indeed a very strange circumstance that the philosophy of the great Ephesian more than adumbrates the two doctrines which have played leading parts, the one in the development of Christian dogma, the other in that of natural science. The former is the conception of the Word {Greek text}[logos] which took its Jewish shape in Alexandria, and its Christian form 4 in that Gospel which is usually referred to an Ephesian source of some five centuries later date; and the latter is that of the struggle for existence. The saying that "strife is father and king of all" {Greek text}[...], ascribed to Heraclitus, would be a not inappropriate motto for the "Origin of Species."

I have referred only to Mr. Sully's article, because his authority is quite sufficient for my purpose. But the consultation of any of the more elaborate histories of Greek philosophy, such as the great work of Zeller, for example, will only bring out the same fact into still more striking prominence. I have professed no "minute acquaintance" with either Indian or Greek philosophy, but I have taken a great deal of pains to secure that such knowledge as I do possess shall be accurate and trustworthy.

In the third place, Mr. Gladstone appears to wish that I should discuss with him the question whether the nebular hypothesis is, or is not, confirmatory of the pentateuchal account of the origin of things. Mr. Gladstone appears to be prepared to enter upon this campaign with a light heart. I confess I am not, and my reason for this backwardness will doubtless surprise Mr. Gladstone. It is that, rather more than a quarter of a century ago (namely, in February 1859), when it was my duty, as President of the Geological Society, to deliver the Anniversary Address, 5 I chose a topic which involved a very careful study of the remarkable cosmogonical speculation, originally promulgated by Immanuel Kant and, subsequently, by Laplace, which is now known as the nebular hypothesis. With the help of such little acquaintance with the principles of physics and astronomy as I had gained, I endeavoured to obtain a clear understanding of this speculation in all its bearings. I am not sure that I succeeded; but of this I am certain, that the problems involved are very difficult, even for those who possess the intellectual discipline requisite for dealing with them. And it was this conviction that led me to express my desire to leave the discussion of the question of the asserted harmony between Genesis and the nebular hypothesis to experts in the appropriate branches of knowledge. And I think my course was a wise one; but as Mr. Gladstone evidently does not understand how there can be any hesitation on my part, unless it arises from a conviction that he is in the right, I may go so far as to set out my difficulties.

They are of two kinds—exegetical and scientific. It appears to me that it is vain to discuss a supposed coincidence between Genesis and science unless we have first settled, on the one hand, what Genesis says, and, on the other hand, what science says.

In the first place, I cannot find any consensus among Biblical scholars as to the meaning of the words, "In the beginning God created the heaven and the earth." Some say that the Hebrew word *bara,* which is translated "create," means "made out of nothing." I venture to object to that rendering, not on the ground of scholarship, but of common sense. Omnipotence itself can surely no more make something "out of" nothing than it can make a triangular circle. What is intended by "made out of nothing" appears to be "caused to come into existence," with the implication that nothing of the same kind previously existed. It is further usually assumed that "the heaven and the earth" means the material substance of the universe. Hence the "Mosaic writer" is taken to imply that where nothing of a material nature previously existed, this substance appeared. That is perfectly conceivable, and therefore no one can deny that it may have happened. But there are other very authoritative critics who say that the ancient Israelite 6 who wrote the passage was not likely to have been capable of such abstract thinking; and that, as a matter of philology, *bara* is commonly used to signify the "fashioning," or "forming," of that which already exists. Now it appears to me that the scientific investigator is wholly incompetent to say anything at all about the first origin of the material universe. The whole power of his organon vanishes when he has to step beyond the chain of natural causes and effects. No form of the nebular hypothesis, that I know of, is necessarily connected with any view of the origination of the nebular substance. Kant's form of it expressly supposes that the nebular material from which one stellar system starts may be nothing but the disintegrated substance of a stellar and planetary system which has just come to an end. Therefore, so far as I can see, one who believes that matter has existed from all eternity has just as much right to hold the nebular hypothesis as one who believes that matter came into existence at a specified epoch. In other words, the nebular hypothesis and the creation hypothesis, up to this point, neither confirm nor oppose one another.

Next, we read in the revisers' version, in which I suppose the ultimate results of critical scholarship to be embodied: "And the earth was waste ['without form,' in the Authorised Version] and void." Most people seem to think that this phraseology intends to imply that the matter out of which the world was to be formed was a veritable "chaos," devoid of law and order. If this interpretation is correct, the nebular hypothesis can have nothing to say to it. The scientific thinker cannot admit the absence of law and order; anywhere or anywhen, in nature. Sometimes law and order are patent and visible to our limited vision; sometimes they are hidden. But every particle of the matter of the most fantastic-looking nebula in the heavens is a realm of law and order in itself; and, that it is so, is the essential condition of the possibility of solar and planetary evolution from the apparent chaos. 7

"Waste" is too vague a term to be worth consideration. "Without form," intelligible enough as a metaphor, if taken literally is absurd; for a material thing existing in space must have a superficies, and if it has a superficies it has a form. The wildest streaks of marestail clouds in the sky, or the most irregular heavenly nebulae, have surely just as much form as a geometrical tetrahedron; and as for "void," how can that be void which is full of matter? As poetry, these lines are vivid and admirable; as a scientific statement, which they must be taken to be if any one is justified in comparing them with another scientific statement, they fail to convey any intelligible conception to my mind.

The account proceeds: "And darkness was upon the face of the deep." So be it; but where, then, is the likeness to the celestial nebulae, of the existence of which we should know nothing unless they shone with a light of their own? "And the spirit of God moved upon the face of the waters." I have met with no form of the nebular hypothesis which involves anything analogous to this process.

I have said enough to explain some of the difficulties which arise in my mind, when I try to ascertain whether there is any foundation for the contention that the statements contained in the first two verses of Genesis are supported by the nebular hypothesis. The result does not appear to me to be exactly favourable to that contention. The nebular hypothesis assumes the existence of matter, having definite properties, as its foundation. Whether such matter was created a few thousand years ago, or whether it has existed through an eternal series of metamorphoses of which our present universe is only the last stage, are alternatives, neither of which is scientifically untenable, and neither scientifically demonstrable. But science knows nothing of any stage in which the universe could be said, in other than a metaphorical and popular sense, to be formless or empty; or in any respect less the seat of law and order than it is now. One might as well talk of a fresh-laid hen's egg being "without form and void," because the chick therein is potential and not actual, as apply such terms to the nebulous mass which contains a potential solar system.

Until some further enlightenment comes to me, then, I confess myself wholly unable to understand the way in which the nebular hypothesis is to be converted into an ally of the "Mosaic writer." 8

But Mr. Gladstone informs us that Professor Dana and Professor Guyot are prepared to prove that the "first or cosmogonical portion of the Proem not only accords with, but teaches, the nebular hypothesis." There is no one to whose authority on geological questions I am more readily disposed to bow than that of my eminent friend Professor Dana. But I am familiar with what he has previously said on this topic in his well-known and standard work, into which, strangely enough, it does not seem to have occurred to Mr. Gladstone to look before he set out upon his present undertaking; and unless Professor Dana's latest contribution (which I have not yet met with) takes up altogether new ground, I am afraid I shall not be able to extricate myself, by its help, from my present difficulties.

It is a very long time since I began to think about the relations between modern scientifically ascertained truths and the cosmogonical speculations of the writer of Genesis; and, as I think that Mr. Gladstone might have been able to put his case with a good deal more force, if he had thought it worth while to consult the last chapter of Professor Dana's admirable "Manual of Geology," so I think he might have been made aware that he was undertaking an enterprise of which he had not counted the cost, if he had chanced upon a discussion of the subject which I published in 1877. 9

Finally, I should like to draw the attention of those who take interest in these topics to the weighty words of one of the most learned and moderate of Biblical critics: 10—

"A propos de cette premiere page de la Bible, on a coutume de nos jours de disserter, a perte de vue, sur l'accord du recit mosaique avec les sciences naturelles; et comme celles-ci tout eloignees qu'elles sont encore de la perfection absolue, ont rendu populaires et en quelque sorte irrefragables un certain nombre de faits generaux ou de theses fondamentales de la cosmologie et de la geologie, c'est le texte sacre qu'on s'evertue a torturer pour le faire concorder avec ces donnees."

In my paper on the "Interpreters of Nature and the Interpreters of Genesis," while freely availing myself of the rights of a scientific critic, I endeavoured to keep the expression of my views well within those bounds of courtesy which are set by self-respect and consideration for others. I am therefore glad to be favoured with Mr. Gladstone's acknowledgment of the success of my efforts. I only wish that I could accept all the products of Mr. Gladstone's gracious appreciation, but there is one about which, as a matter of honesty, I hesitate. In fact, if I had expressed my meaning better than I seem to have done, I doubt if the particular proffer of Mr. Gladstone's thanks would have been made.

To my mind, whatever doctrine professes to be the result of the application of the accepted rules of inductive and deductive logic to its subject-matter; and which accepts, within the limits which it sets to itself, the supremacy of reason, is Science. Whether the subject-matter consists of realities or unrealities, truths or falsehoods, is quite another question. I conceive that ordinary geometry is science, by reason of its method, and I also believe that its axioms, definitions, and conclusions are all true. However, there is a geometry of four dimensions, which I also believe to be science, because its method professes to be strictly scientific. It is true that I cannot conceive four dimensions in space, and therefore, for me, the whole affair is unreal. But I have known men of great intellectual powers who seemed to have no difficulty either in conceiving them, or, at any rate, in imagining how they could conceive them; and, therefore, four-dimensioned geometry comes under my notion of science. So I think astrology is a science, in so far as it professes to reason logically from principles established by just inductive methods. To prevent misunderstanding, perhaps I had better add that I do not believe one whit in astrology; but no more do I believe in Ptolemaic astronomy, or in the catastrophic geology of my youth, although these, in their day, claimed—and, to my mind, rightly claimed—the name of science. If nothing is to be called science but that which is exactly true from beginning to end, I am afraid there is very little science in the world outside mathematics. Among the physical sciences, I do not know that any could claim more than that it is true within certain limits, so narrow that, for the present at any rate, they may be neglected. If such is the case, I do not see where the line is to be drawn between exactly true, partially true, and mainly untrue forms of science. And what I have said about the current theology at the end of my paper [*supra* pp. 160-163] leaves, I think, no doubt as to the category in which I rank it. For all that, I think it would be not only unjust, but almost impertinent, to refuse the name of science to the "Summa" of St. Thomas or to the "Institutes" of Calvin.

In conclusion, I confess that my supposed "unjaded appetite" for the sort of controversy in which it needed not Mr. Gladstone's express declaration to tell us he is far better practised than I am (though probably, without another express declaration, no one would have suspected that his controversial fires are burning low) is already satiated.

In "Elysium" we conduct scientific discussions in a different medium, and we are liable to threatenings of asphyxia in that "atmosphere of contention" in which Mr. Gladstone has been able to live, alert and vigorous beyond the common race of men, as if it were purest mountain air. I trust that he may long continue to seek truth, under the difficult conditions he has chosen for the search, with unabated energy—I had almost said fire—

May age not wither him, nor custom stale His infinite variety.

But Elysium suits my less robust constitution better, and I beg leave to retire thither, not sorry for my experience of the other region—no one should regret experience—but determined not to repeat it, at any rate in reference to the "plea for revelation."

NOTE ON THE PROPER SENSE OF THE "MOSAIC" NARRATIVE OF THE CREATION.

It has been objected to my argument from Leviticus (*suprà* p. 170) that the Hebrew words translated by "creeping things" in Genesis i. 24 and Leviticus xi. 29, are different; namely, "reh-mes" in the former, "sheh-retz" in the latter. The obvious reply to this objection is that the question is not one of words but of the meaning of words. To borrow an illustration from our own language, if "crawling things" had been used by the translators in Genesis and "creeping things" in Leviticus, it would not have been necessarily implied that they intended to denote different groups of animals. "Sheh-retz" is employed in a wider sense than "reh-mes." There are "sheh-retz" of the waters of the earth, of the air, and of the land. Leviticus speaks of land reptiles, among other animals, as "sheh-retz"; Genesis speaks of all creeping land animals, among which land reptiles are necessarily included, as "reh-mes." Our translators, therefore, have given the true sense when they render both "sheh-retz" and "reh-mes" by "creeping things."

Having taken a good deal of trouble to show what Genesis i.-ii. 4 does not mean, in the preceding pages, perhaps it may be well that I should briefly give my opinion as to what it does mean. I conceive that the unknown author of this part of the Hexateuchal compilation believed, and meant his readers to believe, that his words, as they understood them—that is to say, in their ordinary natural sense—conveyed the "actual historical truth." When he says that such and such things happened, I believe him to mean that they actually occurred and not that he imagined or dreamed them; when he says "day," I believe he uses the word in the popular sense; when he says "made" or "created," I believe he means that they came into being by a process analogous to that which the people whom he addressed called "making" or "creating"; and I think that, unless we forget our present knowledge of nature, and, putting ourselves back into the position of a Phoenician or a Chaldaean philosopher, start from his conception of the world, we shall fail to grasp the meaning of the Hebrew writer. We must conceive the earth to be an immovable, more or less flattened, body, with the vault of heaven above, the watery abyss below and around. We must imagine sun, moon, and stars to be "set" in a "firmament" with, or in, which they move; and above which is yet another watery mass. We must consider "light" and "darkness" to be things, the alternation of which constitutes day and night, independently of the existence of sun, moon, and stars. We must further suppose that, as in the case of the story of the deluge, the Hebrew writer was acquainted with a Gentile (probably Chaldaean or Accadian) account of the origin of things, in which he substantially believed, but which he stripped of all its idolatrous associations by substituting "Elohim" for Ea, Anu, Bel, and the like.

From this point of view the first verse strikes the keynote of the whole. In the beginning "Elohim 11 created the heaven and the earth." Heaven and earth were not primitive existences from which the gods proceeded, as the Gentiles taught; on the contrary, the "Powers" preceded and created heaven and earth. Whether by "creation" is meant "causing to be where nothing was before" or "shaping of something which pre-existed," seems to me to be an insoluble question.

As I have pointed out, the second verse has an interesting parallel in Jeremiah iv. 23: "I beheld the earth, and, lo, it was waste and void; and the heavens, and they had no light." I conceive that there is no more allusion to chaos in the one than in the other. The earth-disk lay in its watery envelope, like the yolk of an egg in the *glaire,* and the spirit, or breath, of Elohim stirred the mass. Light was created as a thing by itself; and its antithesis "darkness" as another thing. It was supposed to be the nature of these two to alternate, and a pair of alternations constituted a "day" in the sense of an unit of time.

The next step was, necessarily, the formation of that "firmament," or dome over the earth-disk, which was supposed to support the celestial waters; and in which sun, moon, and stars were conceived to be set, as in a sort of orrery. The earth was still surrounded and covered by the lower waters, but the upper were separated from it by the "firmament," beneath which what we call the air lay. A second alternation of darkness and light marks the lapse of time.

After this, the waters which covered the earth-disk, under the firmament, were drawn away into certain regions, which became seas, while the part laid bare became dry land. In accordance with the notion, universally accepted in antiquity, that moist earth possesses the potentiality of giving rise to living beings, the land, at the command of Elohim, "put forth" all sorts of plants. They are made to appear thus early, not, I apprehend, from any notion that plants are lower in the scale of being than animals (which would seem to be inconsistent with the prevalence of tree worship among ancient people), but rather because animals obviously depend on plants; and because, without crops and harvests, there seemed to be no particular need of heavenly signs for the seasons.

These were provided by the fourth day's work. Light existed already; but now vehicles for the distribution of light, in a special manner and with varying degrees of intensity, were provided. I conceive that the previous alternations of light and darkness were supposed to go on; but that the "light" was strengthened during the daytime by the sun, which, as a source of heat as well as of light, glided up the firmament from the east, and slid down in the west, each day. Very probably each day's sun was supposed to be a new one. And as the light of the day was strengthened by the sun, so the darkness of the night was weakened by the moon, which regularly waxed and waned every month. The stars are, as it were, thrown in. And nothing can more sharply mark the doctrinal purpose of the author, than the manner in which he deals with the heavenly bodies, which the Gentiles identified so closely with their gods, as if they were mere accessories to the almanac.

Animals come next in order of creation, and the general notion of the writer seems to be that they were produced by the medium in which they live; that is to say, the aquatic animals by the waters, and the terrestrial animals by the land. But there was a difficulty about flying things, such as bats, birds, and insects. The cosmogonist seems to have had no conception of "air" as an elemental body. His "elements" are earth and water, and he ignores air as much as he does fire. Birds "fly above the earth in the open firmament" or "on the face of the expanse" of heaven. They are not said to fly through the air. The choice of a generative medium for flying things, therefore, seemed to lie between water and earth; and, if we take into account the conspicuousness of the great flocks of water-birds and the swarms of winged insects, which appear to arise from water, I think the preference of water becomes intelligible. However, I do not put this forward as more than a probable hypothesis. As to the creation of aquatic animals on the fifth, that of land animals on the sixth day, and that of man last of all, I presume the order was determined by the fact that man could hardly receive dominion over the living world before it existed; and that the "cattle" were not wanted until he was about to make his appearance. The other terrestrial animals would naturally be associated with the cattle.

The absurdity of imagining that any conception, analogous to that of a zoological classification, was in the mind of the writer will be apparent, when we consider that the fifth day's work must include the zoologist's *Cetacea, Sirenia,* and seals, 12 all of which are *Mammalia;* all birds, turtles, sea-snakes and, presumably, the fresh water *Reptilia* and *Amphibia;* with the great majority of *Invertebrata.*

The creation of man is announced as a separate act, resulting from a particular resolution of Elohim to "make man in our image, after our likeness." To learn what this remarkable phrase means we must turn to the fifth chapter of Genesis, the work of the same writer. "In the day that Elohim created man, in the likeness of Elohim made he him; male and female created he them; and blessed them and called their name Adam in the day when they were created. And Adam lived an hundred and thirty years and begat *a son* in his own likeness, after his image; and called his name Seth." I find it impossible to read this passage without being convinced that, when the writer says Adam was made in the likeness of Elohim, he means the same sort of likeness as when he says that Seth was begotten in the likeness of Adam. Whence it follows that his conception of Elohim was completely anthropomorphic.

In all this narrative I can discover nothing which differentiates it, in principle, from other ancient cosmogonies, except the rejection of all gods, save the vague, yet anthropomorphic, Elohim, and the assigning to them anteriority and superiority to the world. It is as utterly irreconcilable with the assured truths of modern science, as it is with the account of the origin of man, plants, and animals given by the writer of the second chief constituent of the Hexateuch in the second chapter of Genesis. This extraordinary story starts with the assumption of the existence of a rainless earth, devoid of plants and herbs of the field. The creation of living beings begins with that of a solitary man; the next thing that happens is the laying out of the Garden of Eden, and the causing the growth from its soil of every tree "that is pleasant to the sight and good for food"; the third act is the formation out of the ground of "every beast of the field, and every fowl of the air"; the fourth and last, the manufacture of the first woman from a rib, extracted from Adam, while in a state of anaesthesia.

Yet there are people who not only profess to take this monstrous legend seriously, but who declare it to be reconcilable with the Elohistic account of the creation!

FOOTNOTES:

1
[*The Nineteenth Century*, 1886.]
2
[Both dolphins and dugongs occur in the Red Sea, porpoises and dolphins in the Mediterranean; so that the "Mosaic writer" may have been acquainted with them.]
3
[I said nothing about "the greater number of schools of Greek philosophy," as Mr. Gladstone implies that I did, but expressly spoke of the "founders of Greek philosophy."]
4
[See Heinze, *Die Lehre vom Logos,* p. 9 *et seq.*]
5
[Reprinted in *Lay Sermons, Addresses, and Reviews,* 1870.]
6
["Ancient," doubtless, but his antiquity must not be exaggerated. For example, there is no proof that the "Mosaic" cosmogony was known to the Israelites of Solomon's time.]

7

[When Jeremiah (iv. 23) says, "I beheld the earth, and, lo, it was waste and void," he certainly does not mean to imply that the form of the earth was less definite, or its substance less solid, than before.]

8

[In looking through the delightful volume recently published by the Astronomer-Royal for Ireland, a day or two ago, I find the following remarks on the nebular hypothesis, which I should have been glad to quote in my text if I had known them sooner:—

"Nor can it be ever more than a speculation; it cannot be established by observation, nor can it be proved by calculation. It is merely a conjecture, more or less plausible, but perhaps in some degree, necessarily true, if our present laws of heat, as we understand them, admit of the extreme application here required, and if the present order of things has reigned for sufficient time without the intervention of any influence at present known to us" (*The Story of the Heavens,* p. 506).

Would any prudent advocate base a plea, either for or against revelation, upon the coincidence, or want of coincidence, of the declarations of the latter with the requirements of an hypothesis thus guardedly dealt with by an astronomical expert?]

9

[Lectures on Evolution delivered in New York (American Addresses).]

10

[Reuss, *L'Histoire Sainte et la Loi,* vol. i, p. 275.]

11

[For the sense of the term "Elohim," see the essay entitled "The Evolution of Theology" at the end of this volume.]

12

[Perhaps even hippopotamuses and otters!]

Printed in Great Britain
by Amazon.co.uk, Ltd.,
Marston Gate.